This book belongs to

Welcome to your SPACE ADVENTURE!

Lets pack for your trip!
A SPACESUIT
a ROCKET
and some ANIMAL friends

If you enjoy this product, please check out our offerings

© Kurious Kid

Meet the Planets

Our home is EARTH
We have 7 neighboring planets
We are part of the same SOLAR SYSTEM

SOLAR SYSTEM

Consists of the SUN Planets including the Earth move around it

NEPTUNE

URANUS

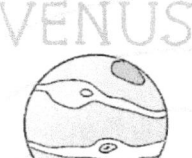
VENUS

JUPITER

MERCURY

SUN

EARTH

PLUTO

MARS

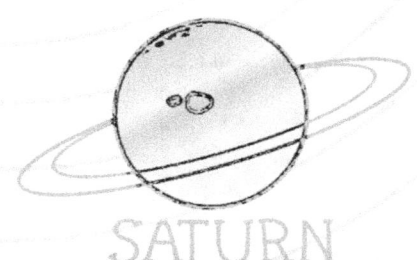
SATURN

MERCURY

The smallest planet
Closest to the SUN
A day on Mercury lasts 59 Earth days
Has no moon

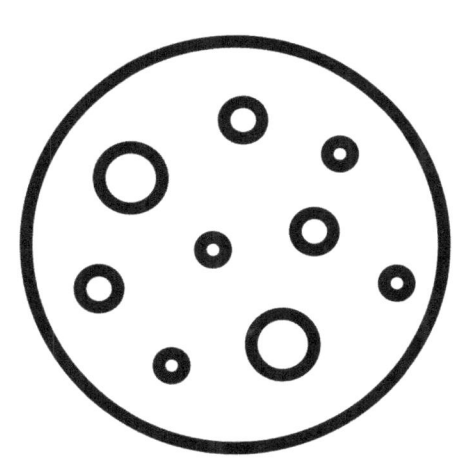

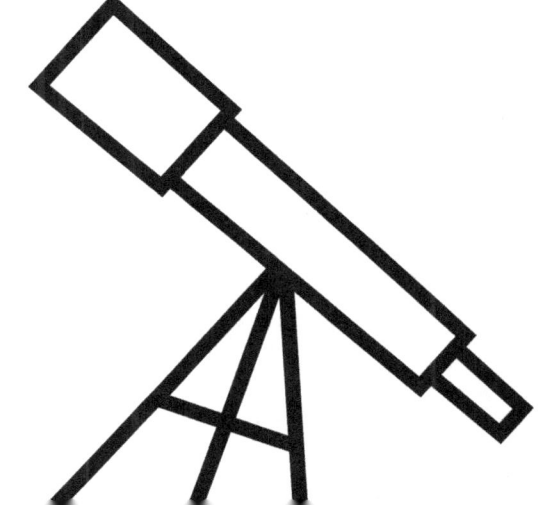

VENUS

The hottest planet
Has volcanoes
A day on Venus lasts 243 Earth days
Has no moon

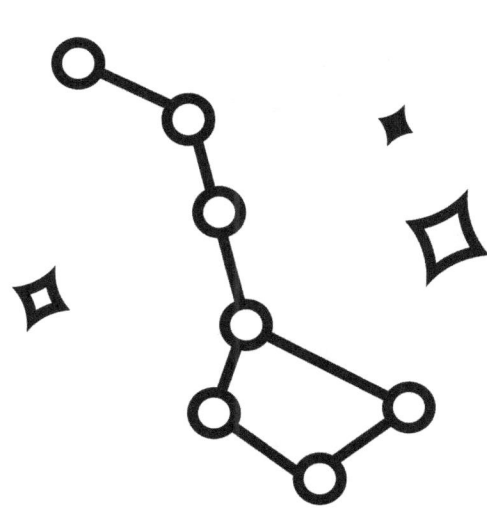

EARTH

**An ocean planet
Supports life
Has 1 moon**

MARS

The red planet
Small & rocky
A day on Mars is
24.6 hours
Has 2 moons

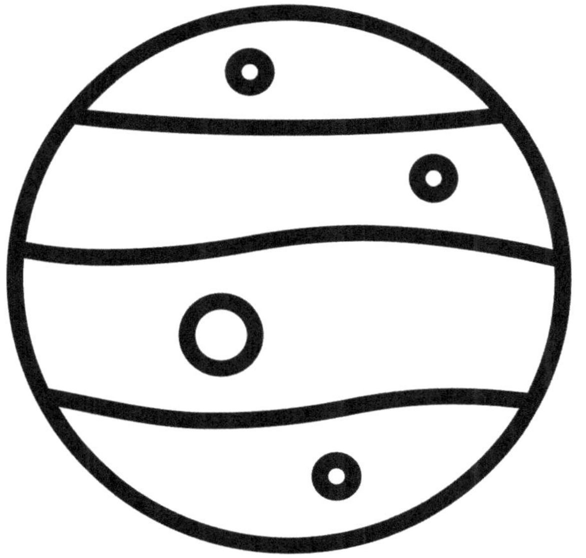

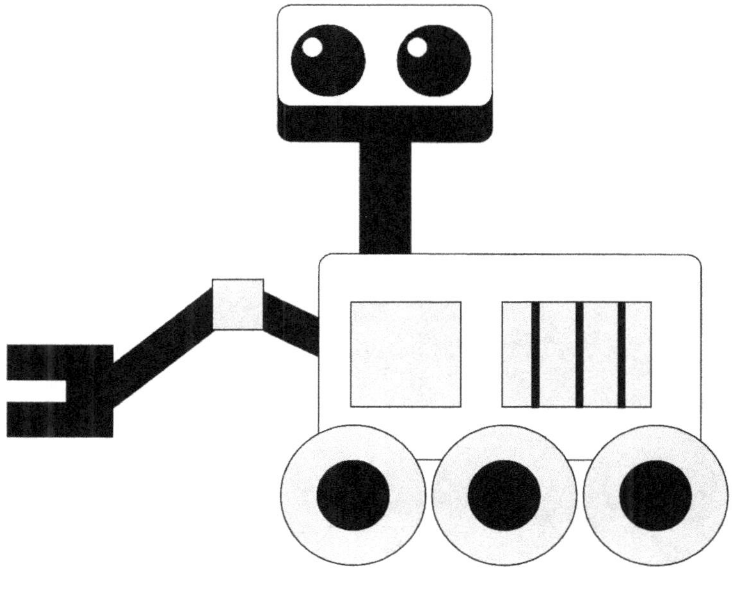

JUPITER

The biggest planet
Is gassy
A day in Jupiter is
just 10 hours
Has 79 known
moons

SATURN

Is gassy
Has beautiful rings
A day on Saturn is only 10.7 hours
Has 53 known moons

URANUS

Is blue
Spins on its side
A day on Uranus is just over 17 hours
Has 27 known moons

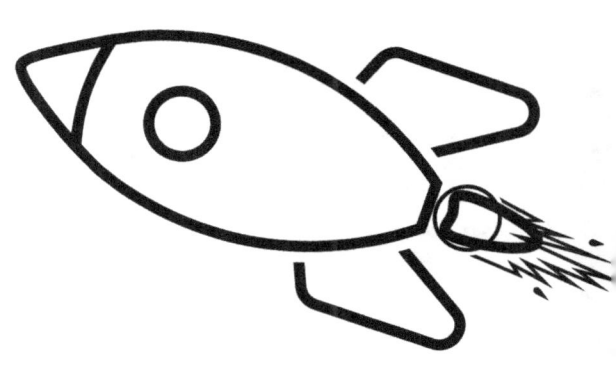

NEPTUNE

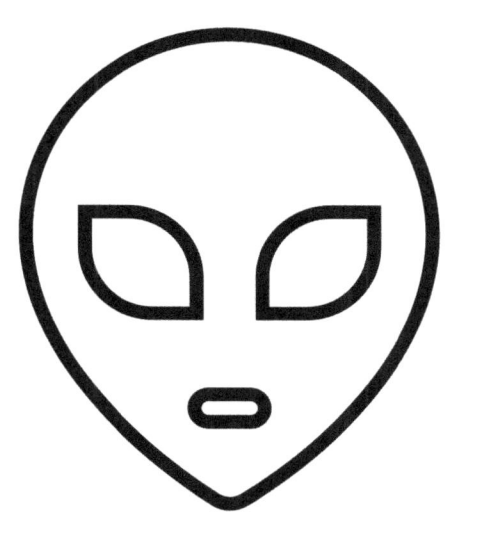

Dark, cold, windy
An ice giant
A day in Neptune is 16 hours
Has 16 known moons

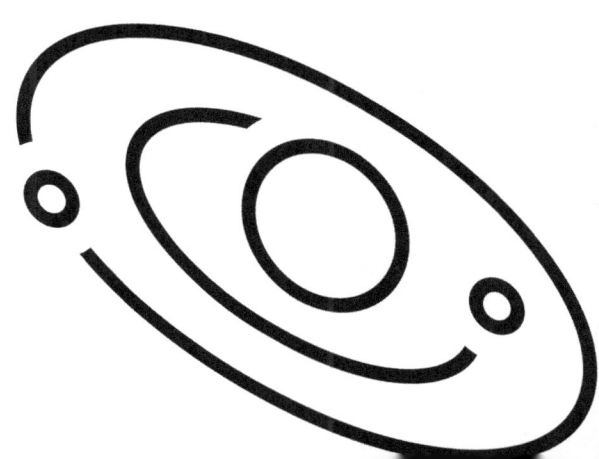

Pluto

**A dwarf planet
Only half the
width of the
United States
Its surface is very
cold**

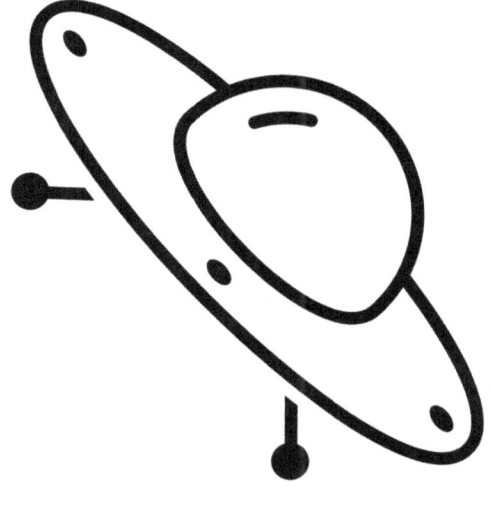

www.ingramcontent.com/pod-product-compliance
Lightning Source LLC
Chambersburg PA
CBHW080911220526
45466CB00011BA/3549